Farm Animals

Chickens on the Farm

by Bizzy Harris

Bullfrog Books

Ideas for Parents and Teachers

Bullfrog Books let children practice reading informational text at the earliest reading levels. Repetition, familiar words, and photo labels support early readers.

Before Reading

- Discuss the cover photo. What does it tell them?
- Look at the picture glossary together. Read and discuss the words.

Read the Book

- "Walk" through the book and look at the photos. Let the child ask questions. Point out the photo labels.
- Read the book to the child, or have him or her read independently.

After Reading

- Prompt the child to think more. Ask: What did you know about chickens before reading this book? What more would you like to learn about them?

Bullfrog Books are published by Jump!
5357 Penn Avenue South
Minneapolis, MN 55419
www.jumplibrary.com

Library of Congress Cataloging-in-Publication Data

Names: Harris, Bizzy, author.
Title: Chickens on the farm / by Bizzy Harris.
Description: Minneapolis, MN: Jump!, Inc., [2021]
Series: Farm animals | Includes index.
Audience: Ages 5–8 | Audience: Grades K–1
Identifiers: LCCN 2020021069 (print)
LCCN 2020021070 (ebook)
ISBN 9781645277019 (hardcover)
ISBN 9781645277026 (ebook)
Subjects: LCSH: Chickens—Juvenile literature.
Livestock—Juvenile literature.
Classification: LCC SF487.5 .H369 2021 (print)
LCC SF487.5 (ebook) | DDC 636.5—dc23
LC record available at https://lccn.loc.gov/2020021069
LC ebook record available at https://lccn.loc.gov/2020021070

Editor: Eliza Leahy
Designer: Molly Ballanger

Photo Credits: PCHT/Shutterstock, cover (left), 22; Valentina_S/Shutterstock, cover (middle); Tsekhmister/Shutterstock, cover (right); lunamarina/Shutterstock, 1; cynoclub/Shutterstock, 3; Roman Demkiv/Shutterstock, 4, 23bl; Modfos/Dreamstime, 5, 23tl; Dima Fadeev/Shutterstock, 6–7, 23tm; AjayTvm/Shutterstock, 8–9, 23br; Oleksandr Lytvynenko/Shutterstock, 10; Jolanda de Jong-Jansen/Shutterstock, 11; BePhumirat/Shutterstock, 12–13; Szasz-Fabian Jozsef/Shutterstock, 14–15, 23tr; John Daniels/Pantheon/SuperStock, 16; Chrisdunham/Dreamstime, 17; Alexandru Ionas Salagean/Dreamstime, 18–19, 23bm; JuneChalida/Shutterstock, 20–21; WildlifeWorld/Shutterstock, 24.

Printed in the United States of America at Corporate Graphics in North Mankato, Minnesota.

Table of Contents

Hens and Chicks

The chickens huddle.
The coop is warm.

Good morning!

The chickens come out of the coop.

tail
comb

A rooster crows.
He has a big tail.
He has a red comb.

This one scratches the dirt.

His claws help.

He looks for bugs to eat.

claw

Some hens are brown.
Some are white.

These have many colors.
See their feathers?

This hen sits on her eggs.

Why?

She keeps them warm.

egg

chick

Peep! Peep!

What is that sound?

Chicks!

They hatched!

Time to leave the coop!

The chicks stay under Mom's wing.

Why?

They stay warm.

They look for food.
They peck at the grass.
Mom stays close.

The farmer feeds them.

Eat up!

Parts of a Chicken

Take a look at the parts of a rooster!

Picture Glossary

coop
A small building or pen used to house chickens or other small animals.

crows
Makes the loud sound of a rooster.

hatched
Broke out of a shell.

huddle
To come together closely in a group.

peck
To strike or pick up something with a beak.

scratches
Digs at with claws.

Index

To Learn More

Finding more information is as easy as 1, 2, 3.

1. Go to www.factsurfer.com
2. Enter "chickensonthefarm" into the search box.
3. Choose your book to see a list of websites.